A

SPACE

EXPLORER'S

THEOLOGICAL

TRAINING GUIDE TO

ALIEN ENCOUNTERS

UNDERSTANDING ALIEN

ENCOUNTERS

FROM A THEOLOGICAL PERSPECTIVE

Michael Richard Craig

"The mere existence of an alien life form from other worlds should never be cause for concern or reason to fear, for it is only a revelation of truth."

"We should always assume there *are* alien forms of life *yet* to discover, understand and explain, while at the same time expecting they *will* be discovered, understood and explained."

"The only truth we need to fear is truth that has been distorted"

THIS MANUAL IS DEDICATED TO EVERYONE WHO LOVES THE TRUTH, WHEREVER IT MAY BE FOUND.

CONTENTS

ACKNOWLEDGMENTS

Special thanks to my wife, Carol, and my sons, Jonathan and Stephen, for their love, inspiration and joy!

Thanks also to Mr. David Kyle, a father of science fiction and also a fellow train passenger, and lover of truth and imagination.

✦

CHAPTER ONE

THE STARTING POINT

Space exploration is not much different from theology in that it is about the discovery of truth. In space exploration, the truth we find may or may not be familiar to us. The truth may surprise, frighten, confuse, and challenge us. Yet, this truth in and of itself is not to be feared.

The discovery of alien life forms, apart from our own, may challenge our theological and religious beliefs, but this challenge should not cause despair. On the other hand, our beliefs should be adjusted and enhanced by it. For, it is the very duty of our beliefs to bring us to a knowledge of and conformation to truth. We should never reject, ignore, or be repelled by it.

This must be understood by those who seek to go where no one else has gone before, and encounter what no one else has experienced before through space exploration. It must also be understood by those who train them for their mission.

Therefore, as a part of ongoing space explorer training, this theological aspect of their preparation must not be ignored, for it is essential in developing the most qualified and prepared individuals for the mission of space exploration.

This manual will provide a summary of some of the more important aspects of theological perspectives and preparation for alien encounters. A more detailed and comprehensive study will be prepared for actual individual and class training.

The basic objective of this manual is to provide a beginning point for individual self-examination in matters related to personal "world-life" views including faith, religion and theology and how a space explorer's personal belief system might impact his or her encounter with alien life forms.

CHAPTER TWO

THE ASSUMPTIONS

1. All life is valuable.

2. There are a variety of alien life forms.

3. There are and always will be alien beings and life forms yet to be discovered.

4. We will discover and encounter some of these life forms.

5. These life forms were created by a force other than themselves.

6. These aliens and other life forms will come and go, live and die.

7. All have a definite beginning and end.

8. All alien life forms need "something" to continue their existence.

9. No life forms are perfect. Each is a mix of strengths and Weaknesses, good and bad.

10. There are two categories of life forms.
    - Lower life forms: not aware of their own existence and unable to communicate.
    - Higher life forms: aware of their own existence and able to communicate.

11. All alien life forms have form and substance.

12. Life forms may exist as one of a combination of the following substances: biological, mechanical, mental, emotion, spirit.

13. Some of these alien beings and life forms will remain undiscovered.

14. These various alien beings and life forms will be at different levels of development in awareness, consciousness and intellect.

15. The alien beings and life forms will function according to a set of known or unknown rules, laws, patterns, morals, and/or ethics.

16. These patterns, rules, laws, morals, and/or ethics will be based upon some foundation of beliefs which can be identified.

17. Those alien life forms that have the capacity and capability will have asked and attempted to answer basic questions about their own existence. i.e. Where did I come from? How was I created? Who or what created me? Why are we here? What is my purpose, meaning and value? And where do I go when I die or cease to exist?

18. Those aliens and life forms we encounter may redefine what is means to be "living", because their existence is so foreign to what we understand.

19. Those who have the capacity and capability to communicate can be communicated with by us.

20. Alien life forms may or may not be dangerous.

21. If we are to have beneficial encounters and relationships we must understand their "worldview".

22. We must always leave our minds open to things we have not considered, known or believed.

23. The existence and first encounter with "lower" alien life forms will have less theological/psychological impact than the existence and first encounter with "higher" life forms on space explorers.

CHAPTER THREE

QUESTIONS FOR CONSIDERATION ABOUT ALIENS

1. How is life defined and determined?

2. Are they "lower" or "higher" alien life forms?

3. What is their form and substance?

4. Can they communicate?

5. Are they dangerous?

6. How do they deal with the death of their own?

7. What does it mean to be a living substance?

8. What is their "world life" view?

9. What do they need to continue to exist or to die?

10. What are their weaknesses and strengths?

11. What can we learn from them and they from us?

12. What do they know of feelings and emotions?

13. How do they function biologically, mentally, emotionally, spiritually?

14. Do they have rituals and what are the meanings?

15. Why do they act the way they do?

16. What senses do they have?

18. If they could change one thing about themselves or their society what would that one thing be?

19. How do they explain meaning, worth and value?

20. How do they function emotionally?

21. What do they know about their own history and existence?

22. Do they believe in a creator?

CHAPTER FOUR

QUESTIONS FOR SPACE EXPLORERS

1. What do I believe?

2. Why do I believe what I believe?

3. What are my beliefs based upon?

4. Is this a solid basis or foundation for my beliefs?

5. How do I know this is a solid foundation or basis for my beliefs?

6. How does the discovery of aliens or other life forms challenge my belief system?

7. How important are my beliefs to me?

8. Am I willing to adjust what I believe to coincide with true reality?

9. Are my beliefs about truth?

10. Can I work, live and play with those whose beliefs are different from mine?

CHAPTER FIVE

BASIC TYPES OF RELIGIOUS BELIEFS

Which of the six categories below best describe your belief? The aliens' belief?

1. Atheism - there is no creator.

2. Agnosticism - the existence or non-existence of a creator cannot be known.

3. Pantheism - the creator is the universe/the universe is the creator.

4. Deism - There is a creator(s) but the creator(s) does(do) not interact with creation.

5. Theism - there is a creator and the creator does interact with the world.

6. Poly-theism - there are many gods.

CHAPTER SIX

RESPONSE OPTIONS TO AN ALIEN WORLD

1.   Adopt - If we like it, make it ours also.

2.   Adapt - If we like it, change it to fit us.

3.   Avoid - If we do not like, it leave it alone.

4.   Advise - If we see a way to help or improve.

5.   Observe - We simply watch without interference.

6.   Ignore - Do nothing. Do not even observe.

7.   Destroy - If it threatens our existence, destroy it.

CHAPTER SEVEN

RULES OF COMMUNICATION WITH ALIEN LIFE FORMS

The purpose of communicating with "higher" alien life forms should always be to establish peaceful relationships and gain clear understanding of one another's world through mutual sharing of knowledge, experience and history.

1.   Recognize as an alien life form.

2.   Determine if communication is possible.

3.   Determine if safe communication is possible.

4.   Determine the means of communication.

5.   Determine the method of communication.

6.   Determine the parties involved in the communication.

7.   Determine what to do with or how to respond to the communication.

8.   Record/Determine the date, time and location of the communication.

CHAPTER EIGHT

STAGES OF COMMUNICATION WITH "HIGHER" ALIEN LIFE FORMS

1.   Phase One: Avoid the following on initial contact

- Judging
- Condemning
- Offending
- Challenging
- Confronting
- Changing

2.   Phase Two: Allow the following after initial contact

- Questioning
- Sharing
- Debating
- Discussing
- Exchanging
- Persuading
- Examining

Note: It is essential to define terms, concepts and actions clearly and in context through the following process:

1.   Listening/observing to what is being communicated.

2.   Explaining how you understand and interpret the communication.

3.   Have them repeat back and clarify your explanation.

CHAPTER NINE

SCORING ALIEN LIFE FORM ENCOUNTERS

In order to establish the characteristics of the alien life form encountered the following scoring grid has been established.

Questions	1	2	3	4	5	6	7	8	9	10
Answers - X mark only										
Individual Answer Score										
Total Score										

1. Alien life for can communicate.

2. Alien life form is not hostile/dangerous.

3. Alien life form is aware of its own existence and asks basic life questions.

4. Alien life form is a familiar life form.

5. Alien life form is similar to us.

6. Alien life form cannot communicate.

7. Alien life form is hostile/dangerous.

8. Alien life form is not aware of its own existence and does not ask basic life questions.

9. Alien life form is unfamiliar life form.

10. Alien life form is not similar to us.

Directions for using the above scoring grid:

1. The first row lists the number for each of the 10 questions listed above.

2. The second row provides a block for marking an X if the question number in the block above applies to the alien life form.

3. The third row provides a block for recording the score for each X marked in the block above it. The value of the X will equal the number in the block directly above it. Example: If an X was marked in block #1 on the second row, then the score of 1 would be put in the block directly under the X on the third row.

4. The total score of all the numbers in the third row should be added together and entered into the single available box at the end of the fourth row.

5. The lower the "total score" indicates a more positive alien encounter outcome. The higher the "total score" indicates a less positive alien encounter outcome.

CHAPTER 10

The following information is provided to illustrate how a person of faith (specifically an ordained minister serving as a chaplain) might have an impact in helping alien encounters be understood from a theological perspective.

THE PURPOSE AND ROLE OF A CHAPLAIN IN SPACE EXPLORATION

- To incorporate into space exploration a system for addressing alien encounters from a theological/religious perspective.

- To promote the free exercise of religion and provide for the religious needs of all command personnel.

- To serve as an advisor to the command on all matters of a theological, religious, moral and ethical nature in the light of the existence of alien life forms.

THE QUALIFICATIONS OF THE CHAPLAIN

Bachelors Degree

Masters Degree in Religion/Theology

Ordination in specific faith group

Two years experience

Direct Commission into United States Space Force

Chaplain Corps/Astronaunt Indoctrination training

THE CHAIN OF COMMAND OF THE CHALAIN CORPS

United States Space Force Chief of Chaplains

Space Fleet Chaplains (North, South, East and West)

Space Fleet Sector Chaplains

Space Station and Space Ship Chaplains

Staff Chaplains

Note: Chaplains are set within the same promotion, rank and pay structure as all other officers but reports to and works directly for the Commanding Officer.

A DEFINITION OF THE CHAPLAIN CORPS

An organization consisting of ordained and trained clergy from various faith groups and religious denominations who meet the requirements and have received a direct commission into the United States Space Force Chaplain Corps.

CHAPTER 11

CHAPLAIN CHARACTER DESCRIPTION FOR THE STORY LINE

The idea is to not make the chaplain too preachy. He is well educated, yet down to earth with a good heart. He has a great but dry sense if humor. He is well respected and his input is sought and valued but not always taken. He makes mistakes but is confident, bold and not easily taken advantage of. He does have a bit of dark side, a struggle that stays deep but constantly tries to unveil itself. He is a mix of Captain Kirk, Magnum PI and Bugs Bunny with a bit of Popeye thrown in the mix.

STORY LINES

Fear

They Always Were

Saving a Ship

Something and Nothing

The Meaning of Death

Loving Aliens

Story Line Dialogue

FEAR

Setting: The chaplain is approached by a fellow officer who wants to meet with the chaplain. They go to the chaplain's stateroom on the spaceship.

Chaplain: Jack, how are you? What's on your mind?

Jack: Chaplain, you know me. You know my faith, but I am ashamed to say I am afraid. I'm terrified!

Chaplain: Jack. What are you afraid of, terrified about?

Jack: I am so afraid of what we will find, what we will see. I've prepared my whole life to be in this very position and I'm afraid. How can my faith handle knowing there is more out there, other life forms, other beings? How can I maintain my beliefs knowing there are others worlds?

Chaplain: Jack it's ok to be afraid. There is nothing wrong with the way you feel. Your faith can handle it. Let me explain something to you. You are really afraid of the truth. Yet, I know you love the truth, just as I do. The problem is, is that you have narrowly defined the truth. You have limited the truth to your own little world. Its boundaries far exceed the limits you have set.

Jack: I'm not sure I completely understand what you mean chaplain.

Chaplain: Jack, the truth is the fabric that holds the universe together. It extends from one end to the other and has been here from the very beginning and will always be here. Wherever we go truth is there and its existence should never make us fear. Truth just is. We have to allow ourselves to open up to what it reveals, relax with what it shows us and then we find peace.

Jack: I'm getting you now. My faith is about believing the truth. Therefore, the revelation of truth should not damage what I believe? My faith can handle it!

Chaplain: Exactly Jack! Don't be afraid of the existence of other worlds, other life forms, of aliens. They do exist. Embrace that truth.

Jack: Chaplain you have helped me make sense of this, but I need to let it sink in a bit, you know. It may be something I'll want to follow up with you.

Chaplain: You bet Jack. Anytime.

THEY ALWAYS WERE

Setting: The chaplain is in a staff meeting with the captain (CO), executive officer (XO) and the department heads of the ship as discussion takes place about an encounter with an alien ship and their crew.

Captain: Chaplain what's your concern.

Chaplain: Something is not right. I just don't think we can trust them.

Captain: Tell us what's on your mind.

Chaplain: There is no doubt they have displayed power unlike anything we have seen. Their intelligence and knowledge is far beyond what we can understand. Yet, their claims about themselves and their origin are very troubling. They have said that they have always existed, that they were the first of all life. They said they have no creator but themselves.

Captain: Go on chaplain. Explain your concern about this.

Chaplain: How can something that already exists have created itself? It would have had to exist before it existed in order to create itself. That is an impossibility!

Executive Officer: So what are you saying? They are lying to us, trying to deceive us?

Chaplain: I am saying, either they are lying to us and know it and are trying to deceive us for some reason or they have no clue and are deluded in their thinking and are not as smart as we think they are. Captain I suggest we keep this possible deception or delusion in mind as we negotiation with them.

Captain: Any thoughts from anyone else?

Executive Officer: Chaplain, why would they lie about this? What could they be hiding?

Chaplain: That is the question! If we can answer that, then we know what we are facing.

SAVING A SHIP

Setting: The chaplain is on the ship's bridge with the captain after having encountered hostile actions from what seemed a peaceful alien ship.

Chaplain: Captain, I beg of you not to do this thing! I think I know why they have reacted this way. Let me go to them one more time.

Captain: No way chaplain! My first priority is to protect this ship. You of all people know that.

Chaplain: I do know that sir, but it's all a misunderstanding. It has to do with their beliefs, their religion, if you will. We seemed to have deeply offended them and they see this as the only way that can respond to what we have done. They feel they have no other choice.

Captain: Chaplain what are you referring to? What have we done to offend them? What religious beliefs?

Chaplain: Captain trust me. Send me to their ship. I take full responsibility for my life.

Captain: Your life is my responsibility. If I send you that may be the end of it indeed!

Chaplain: Captain. At least it is a chance to save them. One last chance. I know I can clear this up. Just send me captain!

Captain: OK Chaplain, but you are not going alone.

SOMETHING OUT OF NOTHING

Setting: Captain and chaplain are looking out of the ship into deep space at the formation of a star.

Captain: Chaplain we are witnessing the birth of a world. Perhaps this is the exact way our earth, our world, came into existence. What an amazing site!

Chaplain: It is amazing! How can seeing this not cause a person to ask the basic questions of life?

Captain: You mean, "What does it all mean?"

Chaplain: That is certainly an important question, Why are we here? Who created us and why? But, I was thinking of another question. One that precedes yours.

Captain: Well, what question? I'm interested to know.

Chaplain: The deeper question is "Why is there something rather than nothing?"

Captain: An interesting question indeed. Well? You have the answer to that? Please tell me (smiling).

Chaplain: When I get the answer to that, you'll be the first to know. But it does raise other questions, questions like yours.

Chaplain: It may be that we will eventually encounter alien beings that have been created by other beings. But that does not make them the ultimate creator, God if you will. There is an important distinction to be made.

Captain: What distinction?

Chaplain: It is one thing to create something from that which already exists and creating something out of nothing. *Ex nihilo* is part of an ancient Latin saying meaning "out of nothing". It goes with the rest of the statement *nihil fit,* meaning nothing comes. The point is that nothing comes from nothing. Once there is something it is easy to create something, comparatively speaking.

Captain: Interesting concept.

Chaplain: We are not a god because we created, built this ship out of stuff that already exists. No, if we had created it out of absolutely nothing, then that would be something to brag about! Knowing this helps us remember that aliens too have their limits. When we meet the alien who can actually create out of absolutely nothing, then we have not met an alien, we have met the true creator. Our creator. I think we will know.

Captain: Chaplain. That gives an enhanced appreciation for what we see happening.

LOVING ALIENS

Setting: The chaplain is on the mess decks with two enlisted crew members. (one male and one female).

E2: Chaplain if I had my way I would immediately and without question kill any alien I saw.

Chaplain: Wow! That seems pretty rough!

E4: I think I might want to do the same thing. They scare me. They are too dangerous.

Chaplain: Well, I guess that is one way to respond.

E2: Why not? They are not humans. They are freaks.

Chaplain: So anything not human is to be killed. Is worthless?

E2: That is right. Why should I care about some freaky space creature?

Chaplain: Well, perhaps you have a point, but let's think about what you are saying. Animals and fish and birds are not human. They have value and you care about them right?

E2: Well, sure, but that is different.

Chaplain: How is it different? You said anything not human should be killed. Animals, birds and fish and aliens are not human, therefore they should all be killed. They are all worthless, without value.

E4: I think you have a good point chaplain.

Chaplain: I think all life have value and worth and should be respected, even if we do not understand it. I am not saying that there is no life, no alien creature that is not dangerous, but that is a different issue than just saying that just because something is not human it should be killed.

E2: All right, all right. I should have kept my mouth shut.

E4: I am always thinking that!

Chaplain: No, no, not at all. These are the issues that have to be addressed. They must be brought up. It takes a thinker, someone who cares to deal with hard things like this.

E4: So we should love all aliens. Love is the answer. (smiling).

Chaplain: Well, love ain't all bad.

THE MEANING OF DEATH

Setting: A battle is taking place between two alien species and the chaplain and crew are observing unable to intervene.

Weapons Officer: Chaplain they are destroying each other. We know barely anything about them and soon they will destroy one another. What was the purpose of their existence in the first place and more, what is the meaning of their death?

Chaplain: Great question David. How you answer that question depends upon several factors.

David: What factors?

Chaplain: If you believe in an ultimate creator who created them, then you can begin to explore their meaning. The meaning for their life and their death. But, on the other hand, if you believe in no ultimate creator, then there is no answer to your questions.

David: Wait a minute. How do you come to that conclusion? Why does not believing in an ultimate creator mean there is no answer?

Chaplain: It seems reasonable that if there is no ultimate creator, then the only other option is all of life, every living being and all of that which exists, is indeed meaningless, without purpose and ultimate value. For where does purpose, meaning and value come from? The best you can say is that everyone determines that for others and for themselves. If I do not think you have value, then to me you don't. If you are not created and are simply a cosmic, meaningless accident, then you have come from no purpose meaning or value, are here for no purpose, meaning or value and are going to no purpose, meaning or value.

David: I see your point, but that is just your opinion. How is it relevant anyway?

Chaplain: Well, if you can give me another option, I am all ears. There are only two. Either we were created by a creator who has always existed and had reason for creating us or you have a no creator and we are just some continually existing stuff with no ultimate meaning. The reason it matters is because depending upon which option you choose, will affect the way you look at all of life, alien life.

David: And? How so?

Chaplain: I may not like you, I may hate you (I really like you, you are a handsome chap) but regardless how I feel about you, that does not change the reality of you having meaning, purpose and value. The creator has given that to you. It is true about you regardless of what I think. Therefore I have an obligation to treat you in such a way that acknowledges that. If no one created you and you are an accident, you have no objective meaning.

David: Chaplain you are something else. You know all this and yet I can beat you in chess.

Chaplain: Well, we all have our weaknesses. But I still expect you to respect my inherent worth and value!

ABOUT THE AUTHOR

Michael Richard Craig is an ordained minister in the Presbyterian Church in America (PCA) and retired after serving as a Chaplain in the U.S. Navy for 20 years. He is a board-certified chaplain in the Association of Professional Chaplains (APC). He has been married to Carol for over 32 years and they have two sons, two daughters-in-law and one grandson.

www.nobodiesartwork.com
nobodiesinc@yahoo.com

The Many Faces of Michael Richard Craig

www.ingramcontent.com/pod-product-compliance
Lightning Source LLC
Chambersburg PA
CBHW071552170526
45166CB00004B/1652